IMAGES
of America

FILLMORE

This aerial view shows Fillmore in 1947. The town of Fillmore and its acres of predominately citrus orchards are nestled between the towering Topatopa Mountains to the north and the Santa Susana Mountains to the south. According to the Historical Census Population for California's counties and the incorporated cities and towns of Ventura County, the population of Fillmore in 1940 was 3,252, which increased to 3,884 by 1950. (Courtesy of the Fillmore Historical Museum.)

On the Cover: A Fillmore Festival street dance was held on Central Avenue on June 22, 1946. World War II was behind the people of the town, and they were ready for a celebration. The street dance concluded the 1946 Frontier Days Festival, sponsored by the Fillmore JayCees. Over 800 couples danced to music provided by the Melody Men, a six-piece dance band from Ventura. A highlight of the dance was the announcement of the winner of a raffle for a 1946 Ford Deluxe Tudor sedan: a gentleman from Los Angeles. (Courtesy of the Fillmore Historical Museum.)

IMAGES
of America

FILLMORE

Carina Monica Montoya
Foreword by Martha Gentry

ISBN 978-1-4671-0918-5

Published by Arcadia Publishing
Charleston, South Carolina

Printed in the United States of America

Library of Congress Control Number: 2022948570

For all general information, please contact Arcadia Publishing:
Telephone 843-853-2070
Fax 843-853-0044
E-mail sales@arcadiapublishing.com
For customer service and orders:
Toll-Free 1-888-313-2665

Visit us on the Internet at www.arcadiapublishing.com

This book is dedicated to the people of Fillmore who embrace its history, growth, beauty, and struggles.

Contents

Foreword

How do towns come into being? In early America, animal trails had turned into Native American trails, which early settlers then followed, and upon reaching a suitable spot, they built a house or a waystation. Towns were planted where two trails crossed or where water was good. The river valley had been populated very early by indigenous tribes, the Chumash and the Aliklik, who settled along the Sespe River and Piru Creek.

The valley changed when the Spanish, then the Mexicans, and finally the Americans discovered it, each leaving their mark. The Spanish founded San Buenaventura Mission in Ventura, naming the Santa Clara River and valley; the Mexicans ranched and farmed grand ranchos; and Americans became homesteaders.

Water and churches were the catalysts of development in the valley. Before the arrival of the Southern Pacific Railroad in 1886, the valley was primarily used for sheep and cattle ranching and growing beans, alfalfa, and barley. After the Sespe Land and Water Company successfully grew oranges on five acres of land in 1888, the valley significantly changed in 10 years, with robust orchards and a variety of crops blanketing the landscape.

The Southern Pacific Railroad coastal route ran between San Francisco and Los Angeles, which included rail through the Santa Clara River valley to Ventura. The first train crossed the Sespe Bridge on January 4, 1887. The rail from Santa Paula to Newhall officially opened on February 8, 1887, and regular service to Los Angeles commenced the following day. The railroad not only transported passengers but also provided farmers faster transport of agricultural products and materials to and from the valley, which triggered a boom in the local citrus industry. The train depot became the nucleus of a growing town with a need for services that included stores, markets, banks, and other businesses situated near the train depot in what became downtown Fillmore.

Carina Monica Montoya, an accomplished author, researcher, and historian, has written another volume to learn from and enjoy its many vintage photographs and stories.

—Martha Gentry
Executive Director and President of the Board of Directors
Fillmore Historical Museum

ACKNOWLEDGMENTS

I am most grateful to Martha Gentry, executive director and president of the Board of Directors of the Fillmore Historical Museum, for her help in providing historical information and photographs for the book and for her editing. I am deeply grateful to Sue Zeider at the Fillmore Historical Museum for her help in researching information, obtaining appropriate photographs, proofreading, and editing of the book. Sue has been instrumental in assuring the accuracy of the historical information contained in the book.

My deep appreciation goes to Ernie and Becky Morales for their help and friendship in providing me significant information on the Mexican community in Fillmore.

My heartfelt appreciation to Daniel A. Haro for his help in obtaining photographs for the last chapter, among many other things, but most of all, for his patience from the beginning of the writing this book to its completion.

Unless otherwise noted, all images appear courtesy of the Fillmore Historical Museum; in chapter seven, unless otherwise noted, all images are courtesy of the author.

INTRODUCTION

The Chumash Indians inhabited much of California's west coastal lands, including the Channel Islands, before the arrival of the Portola expedition on the Pacific coast in 1769. Junipero Serra, a Franciscan priest from Spain, voyaged with the expedition along with other missionaries to spread Christianity in the New World. Twenty-one missions along California's coast from San Diego to Sonoma were built along El Camino Real (the Royal Highway) from 1769 to 1823, with each mission being a day's journey from the next. Padre Juan Crespi, who also voyaged with the Portola expedition, was the only diarist who was present during the full expedition. Crespi recorded the journey of the Portola expedition when it ventured inland, came upon the Santa Clara River valley on August 11, 1769, and set up a campsite in an area that would later become Fillmore.

San Buenaventura Mission was founded by Junipero Serra in 1782. The Chumash provided most of the physical labor in building the mission and serving the mission's daily needs. More significantly, the Chumash built the aqueduct that brought water from the San Buenaventura River to the mission. The aqueduct was seven miles long and constructed of earth and masonry. It was an engineering marvel of its time. All the California missions were self-sufficient and self-sustaining. San Buenaventura Mission was the nucleus of the land that later would become Ventura County. The Spanish brought horses, livestock, and agriculture to the area, making the mission the main producer of agricultural products, wine, meat, candles, and leather goods. Towns began to grow around the mission lands.

Mexico's independence from Spain in 1821 resulted in the secularization of the missions. Mission lands became property of the Mexican government and were divided into large ranchos and granted to individuals, typically former soldiers for their loyal service during the war and individuals with political influence. The area of Fillmore was once part of Rancho Sespe, which totaled 8,881 acres granted to Carlos Antonio Carrillo on May 29, 1834.

The Mexican-American War (1846–1848) ended with the signing of the Treaty of Guadalupe Hidalgo in February 1848. Under the treaty, the United States paid Mexico $15 million for the purchase of California. In 1849, Californians sought statehood, and in 1850, California became the 31st state in the Union. That same year, California designated its first 27 counties. Ventura County today was once part of Santa Barbara County's southern portion until it was split to create Ventura County with the city of Ventura as the county seat.

Before 1888, the Santa Clara River valley was primarily used as range land for cattle and sheep and dry farming of beans, alfalfa, and barley. Properties in the valley were owned by Rancho Camulos, Rancho Santa Paula y Saticoy, and Rancho Sespe. Santa Paula and Saticoy were the only towns in the valley at that time. Later, with news that the Southern Pacific Railroad would be building a line through the area, town plans for Fillmore, Piru, Sespe, and Bardsdale were laid out.

Although Fillmore is said to have been founded in 1887 with the coming of the railroad, it did not become an official town until its incorporation in 1914. In 1888, the Sespe Land and Water Company successfully grew oranges on five acres of land. When others began to successfully grow citrus, it changed the valley's landscape into acres of citrus orchards. When the Southern Pacific Railroad laid tracks through the Santa Clara River valley in 1887, Fillmore began to quickly grow, but the towns of Sespe, Piru, and Bardsdale did not flourish as planned. The rail line stopped short of the town of Piru, and the railroad went north of the Santa Clara River away from the town of Bardsdale. However, the valley became a big producer of citrus and was soon known for its orange and lemon orchards. Agriculture, oil, and oil refining in Fillmore were all dependent on the railroad for transport to reach markets. When oil was discovered in and around the Santa Clara River valley in 1877, many gold miners who came to California during the Gold Rush (1848–1855) became oil workers. More oil was discovered in 1911 on property owned by William Shiells, who became a resident of Ventura County in 1884.

The Mexican American community in Fillmore played a significant role in the agricultural labor force that enabled local farmers to become part of a large-scale industry. Many of the Mexican laborers resided in Rancho Sespe, known as a workers' camp and often called "The Mexican Village." In 1919, a bunkhouse was built, one of three buildings that housed unmarried male workers. The bunkhouse had 14 rooms that were approximately 7 feet by 13 feet. Many of the married workers raised their families at the camp in small cottages. By the 1930s, over 100 families lived at the camp. When Mexican workers built their own small houses, American farmers were better able to sustain a reliable year-round workforce. Unfortunately, it was a time of segregation in America that affected non-whites in general and Mexican, Chinese, Japanese, and Filipino farm laborers in particular in California. It wasn't until 1962, after the United Farm Workers Association was founded, that agricultural workers were given better housing conditions and pay. The Civil Rights Act of 1964, which prohibits discrimination on the basis of race, color, religion, sex, or national origin, enabled minorities to better assimilate and exercise options to improve their lives. Many of the camp's children were able to attain higher education and seek work outside the fields, resulting in many Mexican Americans in Fillmore creating and leaving legacies of significant contributions. Mexican-owned businesses, markets, and stores flourished in the town for decades. Many children of the laborers obtained positions in local law enforcement and as local government officials, occupying city council seats and mayorships. Mexican groups and clubs, such as Los Caballeros and Los Padrinos to name a few, were founded on the spirit of community service, including fundraising to help high school youths by giving scholarships to deserving students.

The first school in the valley was built some time between 1874 and 1875 by a man named Frederick Sprague. Later came grammar schools in Sespe, Fillmore, Willow Grove, San Cayetano, and Montebello, all of which later became the Fillmore Union Grammar School District.

Early settlers obtained religious services through circuit riders who practiced Methodism throughout America. The first Sunday school in the valley was organized in 1873 by families who lived near Cienega, a stage stop on a wagon road.

The Fillmore Post Office was established in 1887, before Fillmore was incorporated. Fillmore's growth began when the railroad laid tracks in the valley in 1887. Central Avenue became the nucleus of the town, where stores and businesses were in close proximity to the train station. A boardinghouse called the Cottage Hotel was located at Main Street and Central Avenue and served as a mail drop before the Fillmore Post Office was built.

Entertainment in Fillmore has always been a big part of the community. In the early years, picnics, barbecues, and relaxation at nearby hot springs were popular. Later, vaudeville and silent movies at the local theater and annual celebrations and festivals, such as the Maypole Festival, became the main sources of the town's entertainment.

The pioneers of Fillmore left a mark that can still be seen today, such as the town's successful agricultural industry. Street names, buildings, historic landmarks, businesses, organizations, and clubs are reminders of the hard work and significant contributions the early pioneers made in building Fillmore to what it is today. Some of the pioneers include Benjamin and Rebecca Saeger Derdorff (1897). Rancher Squire Tietsort and wife Ella (1875) owned property that was later subdivided and became Foothill Drive. Mary, the daughter of Duncan and Julia Cummings

(1890s), was part of the first graduating class from Fillmore High School. James Walker served as Ventura County sheriff from 1892 to 1894. Edward Burson, one of the citrus industry pioneers, planted the third orange grove in Bardsdale. William Chadsey (1880s) lost his daughter, grandson, and son-in-law in the St. Francis Dam Disaster in 1928. William Elder Balcom is the namesake of Balcom Canyon. Kenny Grove Park is named after Cyrus Kenney and his family (1870s). Frank, the son of William Howard (1880s), was a teamster working for the Ventura Refinery in the early 20th century. George Henley (1887) filed a 150-acre mineral claim in Sespe Canyon and quarried brownstone that was shipped throughout the western United States. William Linebarger's son Dallison served as county supervisor of the Fillmore District. Mathew Atmore Jr. (1877) served in the California Cavalry during the Civil War. Bion Dorman (1890) and Wade Moore worked for David Cook in Piru and later in the Torrey oil fields.

Additional pioneers who settled in the Santa Clara River Valley include George King, Jacob Michel and family (1890s), Robert Cruson and family (1874), Joseph Rosenberg and family, William Young and family (1888), Diedrich and Frieda Bartels (1888), Rudolph Haase and family (1889), Gottlieb and Louisa Baldeschweiler (1906), Richard Stewart and family (1888), Thomas Wileman and family (1898), Lena Husser and children (1895), Benjamin Robertson and family (1870s), William Chadsey and wife (1880s), Helen Chadsey McCawley and family (1928), Norman Kellogg and family (1870s), Henry Hiller (1889), Mose Fine and family (1870s), Henry and Laura Udall (1895), James Udall (1895), Owen McGowen and sister Amanda (1875), Alex and Agnes Ritchie (1890), James Allee (1875), R.H. Morey (1888), George Cole and family (1892), James Guthrie and family (1880s), Herbert and Daisy Walker (1898), Thomas Wileman and family (1895), Wilbur Baker (1892), William Brockus Jr. and son Herbert (1895), James and Libbie Culp (1895), C.E. Ingalls (1887), Reuben LeBard and family (1890s), John LeBard and family (1890s), J. Marion Sackett (1884), Alfred Ward and family (1892), George Wengert and family (1898), and Carl Moisling and family (1880s).

Although Fillmore's landscape is still predominately agricultural, the town has grown with modern times and has held its charm and pride for well over a century.

One

A Town Is Born

This 1800s photograph shows Jerome A. Fillmore, who was the general superintendent for the Southern Pacific Railroad at the time when the rail line was laid through the town that would later become Fillmore. Jerome Fillmore supervised the construction of the rail line, and after its completion in 1887, the town was named for him.

This 1940s photograph shows a train engineer standing on a step just outside of the train cab after pulling into the Fillmore station. The rail line was completed in 1887 and enabled a coastal route between San Francisco and Los Angeles. More significantly, it opened Fillmore's agricultural reach in transport and export and greatly helped the development of the town.

On January 4, 1887, the first train crossed over the Sespe Bridge. The Sespe and Bardsdale Bridges allowed for travel over the Sespe River, the Santa Clara River, and streams. The rail line from Santa Paula to Newhall officially opened on February 8, 1887, and began providing regularly scheduled service to Los Angeles the following day. The railroad and bridges cut the travel time in half going into and out of Fillmore.

Harry Peyton and his wife, Haidee Atmore Peyton, are pictured in the mid-1880s. Harry Peyton built bridges and trestles for the Southern Pacific Railroad and was hired to construct the Fillmore Southern Pacific Railroad depot in 1887. He constructed the depot from prebuilt flat panels that were transported to Fillmore on a flatbed railcar. The Fillmore depot is the smallest made with the prebuilt panels. Prior to the depot being built, a boxcar left by Southern Pacific was used as a depot and office. Harry Peyton also built Rancho Sespe Bunkhouse No. 2, which is actually No. 3. Between 1911 to 1919, three bunkhouses were built on Rancho Sespe to house unmarried workers. Rancho Sespe was 4,300 acres used for agriculture and pasture for cattle and horses.

The Fillmore depot is seen in the late 1880s. The depot was the nucleus of the town because it was where new settlers arrived and where agricultural products, materials, and other goods were brought in and shipped out. Businesses and stores were built close to the depot, which soon became downtown Fillmore.

An early 1900s photograph shows a three-mule team and covered wagon in front of the Fillmore train depot. The railroad made it significantly easier for farmers and merchants to ship and receive goods. It also opened the town for passenger travel to and from Fillmore, which greatly contributed to the growth of the town.

Joseph McNab, seen in 1888, was president of Sespe Land and Water Company. The company was incorporated in April 1886 as a subsidiary of the Ventura Water and Improvement Company. One of the stockholders was McNab, who later bought out the other stockholders' interests and moved to Fillmore. Businesses were growing in downtown Fillmore close to the train depot. Below, in August 1888, the Sespe Land and Water Company filed a town plan with Ventura County consisting of 66 acres. The company advertised property for sale in the *Los Angeles Herald*, stating, "The Sespe Land and Water Company will offer, at private sale, Town Lots in Fillmore City, and 3000 acres of the finest Fruit Land in Southern California." The advertisement offered free roundtrip railroad transportation from Los Angeles to Fillmore.

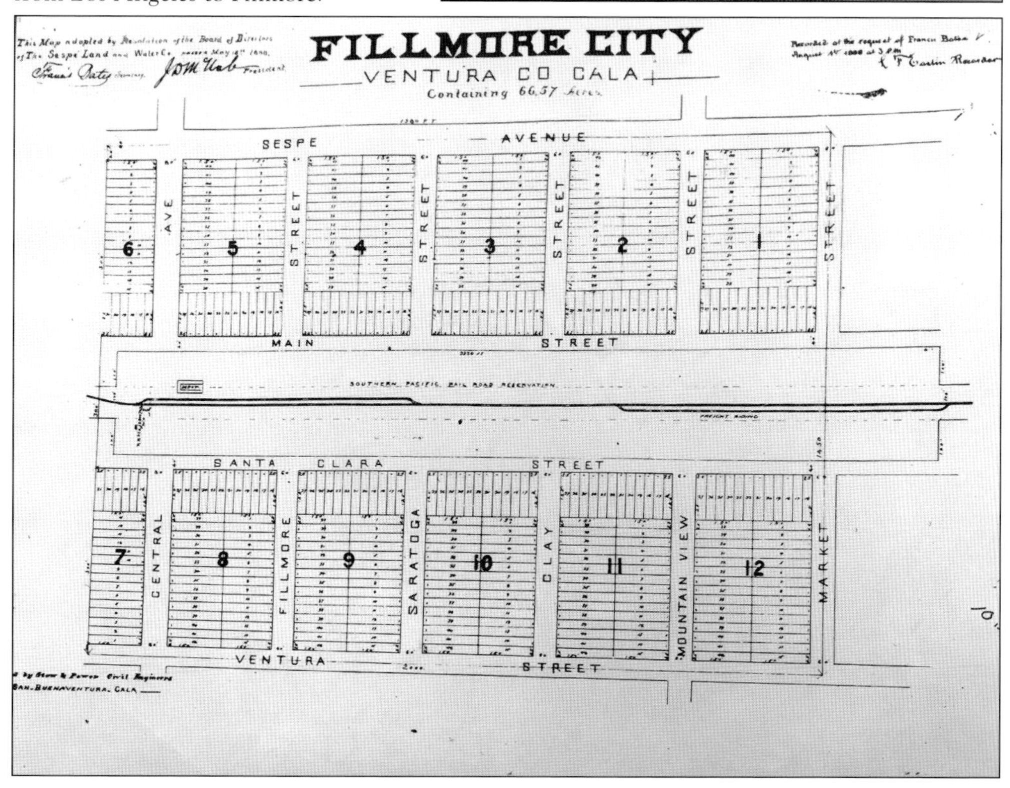

This late-1800s photograph shows Royce Surdam, regarded as Ventura County's first realtor. Surdam came to Ventura County in 1866 and first purchased 1,700 acres of land in Ojai from Thomas R. Bard, a wealthy politician and businessman who assisted in the organization of Ventura County and who was one of the founders of the Unocal Corporation. In 1887, when the railroad came to the valley, Surdam purchased another 1,500 acres from Bard in an area south of the Santa Clara River and west of Grimes Canyon. He laid out a plat for a new town to be called Bardsdale. Below is a copy of Surdam's town plat. When the lots were laid out, Surdam donated land for a school and church. Unfortunately, Bardsdale did not flourish as planned, because the Southern Pacific railroad track was laid on the north side of the river.

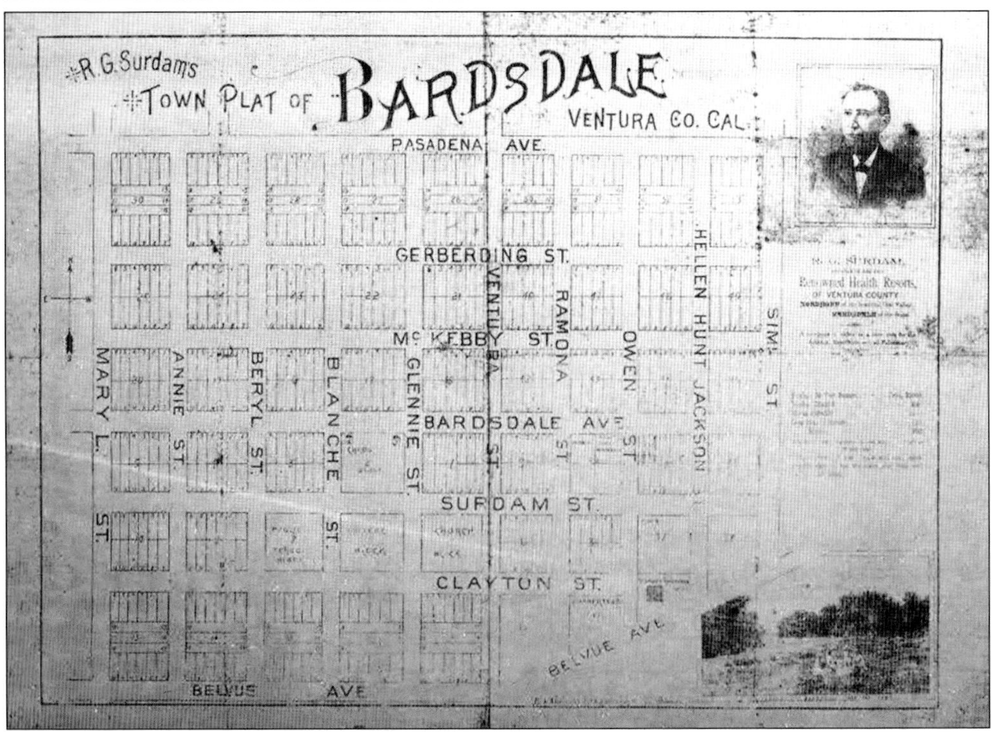

The Sespe train depot is seen in the late 1800s. The owners of Rancho Sespe hoped that Sespe would become the main town of the valley with the coming of the railroad. The town would be built around the Southern Pacific depot. Instead, Fillmore became the fastest growing town in the valley. Below is a map of the laid-out plans for the town of Sespe filed in May 1893 with Ventura County. The land was owned by the Pacific Improvement Company, which promoted the sale of property. Although the plan of the town of Sespe was laid out and water lines were installed, there is no record that the Sespe parcels for sale were heavily advertised as were the towns of Fillmore and Bardsdale.

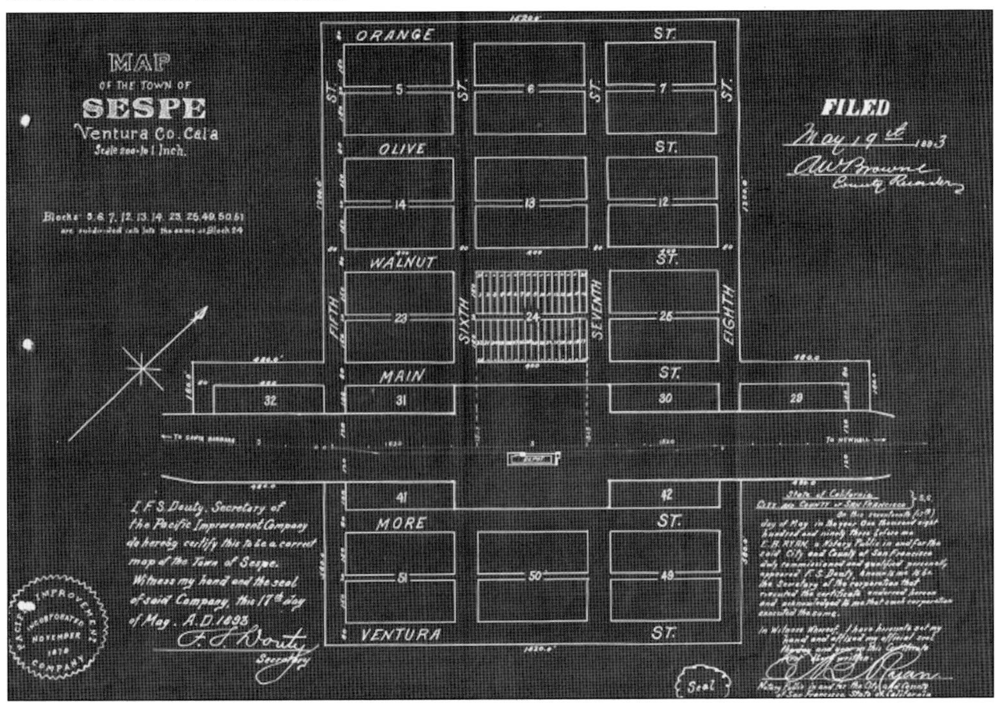

An early-1900s photograph shows Lee Phillips and his wife, Lola, owners of the Sespe Store in Sespe. The store also served as the post office, which moved with each new postmaster. There were seven postmasters from its inception to 1908, when Phillips was appointed as postmaster and moved the post office to his store. Phillips was postmaster until 1932, when the post office was consolidated into the Fillmore Post Office.

In 1907, the Sespe post office also served as a store. The Ventura County Cooperative, which also had stores in Piru and Fillmore, served Sespe in 1912. The store and post office catered to the farmers and ranchers in the Sespe area but also those working in the oil fields and at the brownstone quarry.

This mid-1920s photograph shows the Sespe Post Office, which was once Lee Phillips's store and post office. Although much of the Santa Clara River valley grew with the coming of the railroad, Sespe did not develop along with Fillmore and Santa Paula. The Sespe train depot closed after only a few years, and the post office closed in 1932.

Several workers at Rancho Sespe pose in front of the ranch house in the early 1900s with a bunkhouse at the rear left. Rancho Sespe, composed of six square leagues, was originally granted to Carlos Antonio Carrillo in 1833. It was later purchased by Thomas More and his brothers.

An early-1900s photograph shows Owen Miller, standing at left under the streetlight, and several guests of the Central Hotel. Originally from Pennsylvania, Miller came to Fillmore in 1884. In 1897, Miller became Fillmore's first constable. Other hotels in downtown Fillmore included the Cottage Hotel (mid-1800s) and the Bungalow Inn (1925).

Central Avenue in downtown Fillmore is seen in 1914, after incorporation. Several businesses began to line Central Avenue near the train depot. Early-model cars and horses and buggies travel on the unpaved dirt road. Central Avenue became the main street after a 1903 fire destroyed most of the businesses on Main Street. A bond for paving the streets was up for votes. In the background is a street banner that reads, "Vote for Good Road Bonds."

George Tighe (pictured at right) served as Fillmore's first Southern Pacific train station master in 1887. Tighe owned and operated a general store on Central Avenue in downtown Fillmore. He also owned land and farmed citrus. Although Tighe served as Fillmore's first mayor, he did not support Fillmore's incorporation in 1914. Joel K.L. Schwartz (below) came to Fillmore in 1912 to work as a warehouseman at the train depot. In July 1914, he married a local girl, and he spent the following 20 years working for the railroad. Schwartz was a leading proponent of Fillmore's incorporation. An article in the *Fillmore Herald* stated, "each time the efforts of those who stood for the principals of progress have been frustrated by some move or another, so that this will be the first time that the future welfare of the community was ever really entrusted to the hands and minds of the inhabitants hereof to decide."

Richard Stephens came to Fillmore in 1895, worked as a clerk in James Duncan's store, and soon became a partner in the business. After Duncan's death, Stephens became the sole partner of the business and built a new store in downtown Fillmore. Below, in 1911, several people pose in front of Stephens's new store, decorated with flags and banners, at the corner of Main Street and Central Avenue. It was a celebration of the store's grand opening. Stephens also became Fillmore's sixth postmaster. Stephens's general merchandise store was one of the earliest permanent buildings on Main Street, built before Fillmore's incorporation. Most businesses were established in the downtown area on Central Avenue and on Main Street. The town of Fillmore was blossoming along with its growth in agriculture.

An early-1900s photograph shows one of several blacksmith businesses in Fillmore. This blacksmith and horseshoeing business was owned by William Froehlich (left). The business fabricated objects from iron by hot and cold forging on an anvil. Blacksmiths who specialized in forging horseshoes were called farriers. In the late 1800s and early 1900s, most farmers used mules and horses not only for farm work but also for transportation. Below is M.B. Brinley Horse Shoeing, another blacksmith business in Fillmore. Owner M.B. Brinley (left) and two workers are pictured in front of the shop.

In 1913, proprietor H.S. Robbins (right) and a worker pose in front of the Fillmore Saddlery, which sold stable supplies in downtown Fillmore. A saddlery makes or sells tack, such as saddles, harnesses, and bridles. Below, Fillmore Stables became the oil field headquarters in Fillmore. There were several stables established in Fillmore, such as Jack Trotter's Livery Stable in 1899, Jack Casner's Stables in 1904, Billy Moore's stable in the early 1900s, Bill Elkins Horse Rental, L.L. Warren's Livery Stable in 1915, and Owen Miller's packtrain rental, to name a few.

Dr. Ira Hinckley's drugstore is pictured in the early 1900s. Hinckley was one of the town's earliest dentists and pharmacists. He is pictured at right with Catherine Hinckley (left, behind the counter) and patrons. He established a drugstore at the corner of Central Avenue and Ventura Street that sold medicines, tonics, toiletries, and candy. Below, the drugstore is pictured showing items on display for sale. Ira Hinckley's 1879 diploma from the University of California at Berkeley, a telephone on the back wall, and advertisements for baby food and ice cream can be seen. Ira was also a violinist and a member of the Fillmore Masonic Temple, where he served as worshipful master. The 1930 city directory listed him as secretary of the Masonic lodge.

A 1920 photograph shows the Hardison Sanitary Dairy, founded in 1916 by Elvira and Clifford Hardison. Its name meant the cows were certified as healthy and the raw milk handled according to regulations. The dairy thrived for many years in Fillmore. In 1989, it was declared a county landmark.

Buckman & Co. General Merchandise store is pictured on Central Avenue in downtown Fillmore with workers standing in front and customers arriving in horse-drawn buggies. Flags and banners decorate the front of the store in celebration of its grand opening. It was one of many stores sprouting up in the early 1900s in response to a growing town and a need for services.

Pictured in the late 1800s is one of Fillmore's first automobile garages, Turner & Bookman Garage, advertising the sale of Goodrich tires. When the automotive industry in the United States began in the 1890s and automobiles and trucks began to replace horsepower for transportation and hauling, the need for automotive repair services began to grow. Below, three unidentified workers from the Fillmore Tire Shop pose in front of the business in the early 1930s. The tire store was located on Central Avenue in downtown Fillmore.

A family and other patrons are pictured in front of Fillmore's first bank, called Fillmore State Bank. The bank was incorporated in 1905, before Fillmore was incorporated. Its first location was on the southeast corner of Central Avenue and Santa Clara Street. The bank shared the building with the Masonic hall, which leased the second floor, and the bank occupied the ground floor. Later, both the bank and the Masons constructed their own separate buildings. The bank's new building was on the northeast corner of Main Street and Central Avenue. The Masons built a three-story hall on the northeast corner of Central and Sespe Avenues that was unfortunately destroyed in the 1994 Northridge earthquake. Below, several men serving the Fillmore Volunteer Fire Department pose in front of the second Fillmore State Bank building in the mid-1920s. The bank later became the Bank of Italy, which then changed its name to Bank of America. The building was later bought by the Bank of A. Levy.

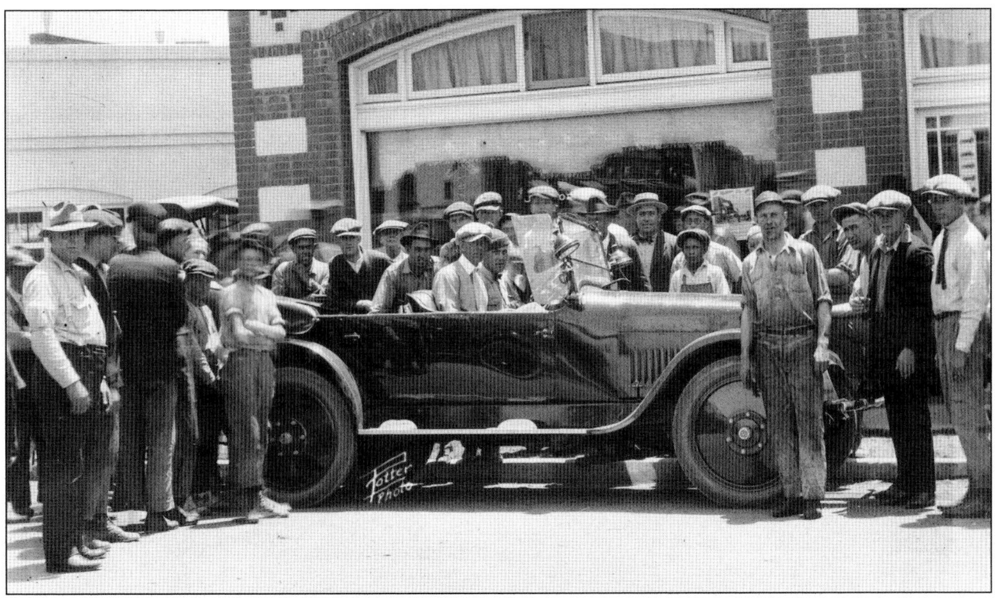

Several people pose for this photograph, taken in front of John Opsahl's Studebaker dealership in Fillmore in the early 1920s. It was the beginning of the era of horseless carriages in Ventura County. Automobiles replaced the stagecoach and were a better and faster means to transport people and things, which in time caused the Southern Pacific's passenger and freight trains to cease running through the valley.

Fillmore's train depot was once the center of the community, but as automobile and truck traffic became more common, train service significantly declined. The last regularly scheduled passenger train came through Fillmore on January 13, 1935. In 1974, Southern Pacific decided to close the Fillmore depot, and eventually, the track between Piru and Newhall was removed.

Dozens of cars are parked along Central Avenue in this 1940s photograph. Central Avenue was the main street where stores and business were located in the downtown area. The photograph was taken looking south down Central Avenue from the top floor of the Masonic building.

A 1945 photograph looks down the east side of Central Avenue from Sespe Avenue. Three men are gathered in front of Security First National Bank, which was the original site of Farmers & Merchants Bank. Two women are seen walking by Patterson Hardware, which was established in 1919 and is still operating today.

Two

Agriculture and Oil Boom

A 1910 photograph shows an orange-picking crew in Fillmore. Citrus picking by hand was very labor intensive. Workers can be seen on ladders, and workers with full fruit-picking bags crate the citrus in the orchards. The crates were then loaded onto a waiting horse-cart to be transported to the packinghouse.

Pictured is Ada Stone Morey in 1915 plowing her ranch with the help of a three-mule team. Mules were ideal for plow pulling, but an iron fist in a velvet glove is the way to dominate the mule. If the mule is allowed to be disobedient while harnessed, it will dominate, and the farmer will end up hoeing the field by hand.

In the 1920s, Floyd Legan Sr. works his ranch with a six-mule plow team. Mules are better adapted to work in hot weather than horses and are more willing to do hard work. When plowing is completed, a mule will still have enough energy to haul a full wagon.

An early-1900s photograph shows workers gathering lima beans for threshing in Bardsdale. Threshing loosens the edible part of grains or other crops from the straw after reaping. One man drives the two-mule team, while two workers with pitchforks load the wagon. The valley primarily engaged in dry-land farming before irrigation.

A crew feeds a threshing machine with harvested peas in 1891. Peas are typically threshed by machine to avoid breaking the seed. When a small quantity of peas needs threshing, a flail can be used. A flail is a threshing tool that has a wooden staff with a short, heavy swing stick.

An early-1900s photograph shows several men and women at work packaging harvested walnuts in crates and burlap bags. Walnuts were first planted in California in the 1800s. Before citrus was grown in the valley, walnuts were farmed along with dry farming crops such as barley, beans, and grains.

Itinerant workers, including children, use a six-mule team and a two-mule team to harvest apricots in 1891. Apricots are typically harvested during the hot summer months from late June through August. By 1935, apricots were one of the most farmed stone fruits and were mostly dried or canned.

In 1910, mostly women and children are ready to prepare apricots for sun-drying. The work was laborious and involved sorting the fruit by size and color, cleaning, and pitting before it could be laid out to sun dry. By the early 1900s, the commercial dried fruit industry was booming in California.

Apricots dry at the E.B. Turner Ranch in Sespe in 1911. Several wood flats of drying apricots can be seen at the left. The growing of apricots in California dates to the founding of the San Buenaventura Mission, when the Spanish missionaries brought the fruit to California.

J.F. McIntyre's apiary (also called a bee yard) is pictured on the Sespe Ranch in 1890. McIntyre's rural apiary was a year-round set-up where honey was produced from the surrounding citrus and wildflowers.

Mel Phillips's beehives is seen in Sespe around the late 1920s. Phillips's apiary was also set up in a rural location surrounded by citrus orchards because the color and scent of the citrus flowers attracts bees.

Several citrus pickers and two women and children pose for this early-1900s photograph. The filled orange crates at right are for the Fillmore Citrus Association. The citrus pickers are wearing fruit-picking bags that have shoulder straps to hold them at waist level and hang in front of the picker. The fruit is deposited into the bag after it is picked.

A late-1800s photograph shows citrus pickers in an orange orchard. Pickers are seen on ladders to reach fruit at the top of the trees, while others work on the ground picking the lower-growing oranges. Citrus was picked by hand and crated in the orchards. Each orange was clipped, dropped into the fruit bag, and then put in a wooden field box to be taken to the packinghouse.

Sespe Ranch bunkhouse No. 1 was located off Telegraph Road and housed the ranch's unmarried male workers during the years from 1910 to 1940. It was one of the earliest facilities in the area to include indoor plumbing, cooking facilities, and electricity. Keith Spalding commissioned the design and construction of the bunkhouse. His wife, Eudora, owned the ranch.

In this 1950s photograph, a small child looks at a dog sitting on the porch of one of the cottages that housed farmworkers with families. A small front and backyard made the small, brown wooden structure with a pitched roof homey. In later years, the cottages were repainted white.

A 1910 photograph shows an unidentified husband and wife crossing the Santa Clara Bridge with a four-horse team pulling their wagonload of oranges bound for a Fillmore packinghouse. Another wagon of fruit headed to a Fillmore packinghouse can be seen following in the distance.

In 1910, a citrus farmer and his mule team haul a wagonload of full orange crates to the Fillmore Citrus Association packinghouse. The California orange boom in the early 1900s made the Santa Clara River valley one of the main producers of citrus in the state.

Pictured in 1921, this truck advertises "First Direct Shipment 'Sunkist' Oranges & Lemons Los Angeles to London." Truck transport expedited delivery of citrus outside of Fillmore to larger cities and abroad faster than train transport, resulting in the closure of the train depots in the valley. Trucks were able to deliver shipments directly to the harbor and airport for transport.

Four packinghouse workers move full crates for transport at the Fillmore Lemon Association in the 1930s. The crates were transported by both train and truck. Brand labels under the Fillmore Lemon Association included Alamo, Selva, Sespe, Wayno, and All Year, to name a few.

Employees of the Fillmore Citrus Association packinghouse pose for the 1913 photograph above. The Santa Clara River valley owes its growth to agriculture. The Fillmore Citrus Association was formed in 1897, and David Felsenthal was president. The association purchased property in 1899 at Sespe Avenue and A Street and built its packinghouse. Below, several workers pack the sorted and cleaned oranges at the Fillmore Citrus Association packinghouse in 1914.

The 1930s photograph above shows the Pure Gold packinghouse, part of the Fillmore Cooperative Orange & Lemon Association, a member of Mutual Orange Distributors (MOD). Below, MOD packinghouse employees pose for a 1930 photograph. Packinghouses are the center of the citrus industry. They connect the orchards to the consumer around the country and abroad. Historically, women have been the majority in packinghouse labor. At the packinghouse, fruits are washed, sorted by size, and assigned a number to indicate how many of a certain size could fit into a crate, then the fruit is graded as being the best, average, or below average.

This 1930s Airship brand is one of the many labels under the Fillmore Citrus Association. The marketing of citrus was genius. The bright oranges, yellows, and greens used in the labels depicted fresh fruit grown in a lush green landscape and signified fast delivery of fruit from the orchard to markets around the world.

This Cycle Brand label shows a cyclist riding along a country road overlooking acres of green citrus orchards. The Sunkist Cooperative California Fruit Growers Exchange focused on health, happiness, prosperity, and respectability. Labels depicted a healthy lifestyle eating fresh oranges and drinking fresh orange juice. According to historian Kevin Starr, by 1914, "consumption of oranges by Americans had increased 79.6 percent, from next-to-no-oranges in 1885."

The Fillmore Citrus Association, seen in the 1930s, was formed in 1897. The citrus boom in the valley created a dire need for packinghouses. Packinghouses created jobs that included picking oranges, loading field boxes onto wagons and trucks to bring the fruit to the packinghouse, unloading fruit from the wagon or truck at the packinghouse, washing fruit, grading for size and quality, packing fruit into boxes or crates, and loading onto trucks or train cars for market.

Workers inside the Fillmore Citrus Association packinghouse are seen in 1906. Female workers pack washed and graded citrus into crates for shipment. Five male workers carry and move filled crates to the bins. The women stand at their stations, where washed and graded fruit is placed into bins ready to be crated.

A 1918 photograph shows male workers at the Fillmore Lemon Association packinghouse. Washed lemons are placed in large flats to be graded and sorted. Here, the workers sort the lemons according to size, color, shape, and defects. The number of workers is evidence of the amount of labor required to process the citrus for market.

Pictured in 1917 is the Fremlin and Walker packinghouse, a locally owned packinghouse on A Street and Telegraph Road. After the Fillmore Citrus Association was formed in 1897, other locally owned packinghouses were established. Two mule teams pulling wagons are being loaded with full citrus crates ready for market.

The Fillmore Citrus Association packinghouse is seen in the 1930s next to the Southern Pacific rail line for easy transport. At left, several boxes of citrus are stacked on a lifting crate to be loaded onto the train. The train had refrigerated railcars, also called "reefers," which allowed large quantities of perishable produce to be transported from farms to cities. The first refrigeration transport was on trains. Later, packinghouses expanded to include cold storage facilities where fruit could be stored for long periods.

This 1915 photograph shows the Sparr Fruit Company packinghouse. Riverside, California-based Sparr was one of the earliest packinghouses in Fillmore. It was located on the southwest corner of Main Street and Central Avenue. The packinghouse was destroyed by fire in 1913 but was rebuilt. The building caught fire again in 1970, long after Sparr had closed for business.

A Rancho Sespe agricultural field is seen in 1920. A worker operates a tillage machine, while two men, possibly foremen, inspect the machine. Tilling the soil, especially after a harvest when cultivation disturbs the soil surface, helps create a good seedbed, loosens the soil for seedlings to grow more easily, allows air exchange, and helps water penetrate more rapidly.

Smudge pots are ready to be used in 1955. A disastrous freeze in 1913 destroyed a whole crop. Since then, smudge pots were used for over 70 years in California's vineyards and citrus groves. Smudge pots were typically containers with crude oil burning in the bottom to provide protection against frost.

In 1915, laborers fumigate citrus trees in Fillmore. Workers entered a tented tree with a lantern for light and ignited a pot containing a pesticide that rendered smoke. The smoke would kill bugs but did not harm the tree. Fumigation was ongoing to prevent citrus pests and diseases that threatened crops.

William Shiells leased extensive land for oil wells, and several rigs are seen dotting the hills. In 1884, William and his brother, James, purchased 1,200 acres south of Fillmore off what is now Guiberson Road. The brothers discovered oil on the property. By the 1900s, oil seekers came to the Santa Clara valley for exploration. The Shiellses leased the property to the Montebello Oil Company in 1910. The land became one of the most productive oil leases in the area. Below, Fillmore's oil refinery is seen in 1920. In 1924, the Ventura Refinery became the Fillmore Texaco Refinery. In 1928, the refinery was known as Fillmore Works. The site was converted to a crude oil pumping and transfer station in 1952.

The St. Francis Dam collapse, which occurred on March 14, 1928, was the worst man-made disaster recorded in Southern California at the time. It took the lives of many and destroyed homes, structures, roads, and crops. Above, residents gathered on a section of a remaining road after the flood passed through Fillmore. The deluge uprooted trees and left debris piled along the Santa Clara riverbank. When Fillmore police chief Earl Hume received the warning of imminent danger, he set out to warn residents who lived close to the river, but by the time he reached the east limits of Fillmore, the flood had hit with a 40-foot wave. Another flood in 1938 destroyed the north approach of the Santa Clara River bridge. A two-passenger cable seat was rigged for people to cross the river at their own peril; the below photograph shows two men crossing the river on the seat.

Three

Mexican American Experience in Fillmore

A big "F" sits on a hill overlooking the city. The letter was put up on May 10, 1936, and has since become a landmark of the town. The letter was built by Frank Morales. Fillmore High School freshmen voluntarily clean the letter each May using 50 sacks of lime. The "F" is 50 yards long and 30 yards at its widest. The letter can be seen from the neighboring town of Santa Paula, and on a clear day, the ocean can be seen from the hill.

Frank S. Morales (at left), along with his cousin Leonard Riesgo and a friend, Mike Sanchez, discussed Santa Paula's letters "SP" sitting prominently in the Santa Paula Hills. They agreed that Fillmore should also have a letter. After scouting the hills overlooking Fillmore, they found the perfect spot. They brought tools to the area and began building the "F," which took two days to complete. Below, Frank is pictured in the early 1930s with his wife, Georgia Alamillo Morales, originally from Globe, Arizona. They married in 1933 and settled in Fillmore, where Frank became an active resident.

A 1940s photograph shows farmworkers' children beneath a US Department of Agriculture sign. Designated farmworker communities were established by the US Department of Agriculture Farm Security Administration (FSA). The FSA was created in 1937 to address rural poverty during the Great Depression in the United States through a photography program that documented the challenges of rural poverty and poor living conditions. The program promoted rural rehabilitation by improving the lifestyle of poor farmers and migrant laborers. Later, the FSA was combined with the Resettlement Administration (RA) to address the need for relief camps for migratory workers. The RA built 95 camps in California that provided a place to live with running water and other amenities. Aside from the FSA and RA, some large farms provided housing for workers, such as at Rancho Sespe. Below, a large family lives in a one-room bungalow for migratory farmworkers provided by the FSA/RA efforts. (Both, courtesy of Library of Congress.)

Lloyd Felix is pictured at left in 1986 at the labor camp owned by the Villasenor family in Fillmore. Lloyd worked as an agricultural laborer and lived at the labor camp for more than 20 years until his death. Below are several small brown houses provided for agricultural workers at Rancho Sespe. Each house provided running water and a small plot to plant a vegetable or flower garden. The small houses were later painted white. The labor camp was called Sespe Village and was also known as the Mexican Village. The camp accommodated 140 families. Farmers who provided quarters for laborers believed that it would foster long-term benefits by having a more reliable workforce. (Left, courtesy of Ellen Frankenstein.)

Several Mexican laborers pick crops in the 1930s. Farm laborers are some of the lowest paid workers in the country and are exposed to some of the most hazardous working conditions, such as pesticides, heat, and lack of shade and clean drinking water, particularly before laws mandated better working conditions and pay for farm laborers. Below, Mexican laborers harvest sugar beets. Harvesting was more labor intensive in the early years, because fruits and vegetable crops were mostly picked by hand. Today, many types of crops still require hand picking. Before citrus and avocado orchards became the landscape of Fillmore, dry farming of beans and grains, and sugar beets around other areas of the county, were the main crops of the valley. An agricultural labor workforce was needed to harvest crops, especially when the valley's citrus boom transformed Fillmore's farming profile, and as the workforce grew, the need for laborer housing grew. The Mexican agricultural labor workforce was the backbone in the success of Fillmore's growing agricultural industry. (Both, courtesy of Library of Congress.)

Pictured is a mixed crew of citrus pickers. Filipino and Japanese laborers were widely used in California's agricultural fields beginning in the early 1900s. America's giant agribusinesses targeted poor islands in the Philippines and recruited single young men between the ages of 18 and 30 to work in America, promising high wages and education opportunities, both of which were not easily attainable before the Civil Rights Act of 1964.

Built in 1916, the Rancho Sespe packinghouse was located north of Highway 126 and Telegraph Road and ran adjacent to the Southern Pacific Railroad. The packinghouse was built under the direction of W.H. Fleet, the Rancho Sespe manager. The facility processed over 4,000 railroad cars of fruit each year during the peak of citrus production.

It can be said that packinghouses are the hubs of the agricultural industry because they connect the farm to the world. Mexican workers, typically females, worked in packinghouses sorting and packaging product. Historically, women have dominated the packinghouse labor since the late 1800s. In this image, Irene Casas inspects lemons for processing at the Lemon House. (Courtesy of Ellen Frankenstein.)

Processing involves a series of treatments between the harvesting and final packaging of any crop, such as washing, drying, sorting, grading, and other procedures depending on the crop. Here, Angie Cuellar is pictured at the Lemon House. Lemons can be seen in different bins after they have been washed, sorted, and graded for appearance in color, shape, and size. (Courtesy of Ellen Frankenstein.)

Pictured above is the Fillmore Citrus Association Mexican Band, established in the early 1920s by Frank Erskine. Erskine was the manager of the Fillmore Citrus Association. The band was comprised of the association's packinghouse and field laborers who had musical interests. The band traveled to perform at events, parades, and celebrations. Erskine was the musical director, and Manuel Lucero conducted. Below, another Mexican musical group in Fillmore that was popular in the 1950s was Los Latinos, made up of eight musicians. The band had a full array of instruments that included the guiro, maracas, electric guitars, trumpets, and drums.

Fillmore's annual May festival and Frontier Days celebrations included beauty pageants and offered activities that were often held on school grounds, at Kenney Grove Park, and on Central Avenue. Music, singing, speaking, winding the maypole, barbecue, and a parade attracted people from all around the area, including neighboring towns and communities such as Colonia in Oxnard. Above, the Colonia Mexican Fillmore float was an entry in the 1928 Fillmore Festival parade. The young children are wearing white dresses. The young woman standing at left is holding the US flag and wearing a striped skirt and hat to symbolize the flag, and the other young woman is wearing a traditional Mexican costume and holding the Mexican flag. Colonia was laid out by the Colonia Land Improvement Company to house workers near the sugar factory and sugar beet fields. Labor leader Cesar Chavez was once a resident. Below, a photograph from September 16, 1945, shows the queen (to the right of the banner) and queen candidates in Fillmore.

Above is Rancho Sespe and its bunkhouses. There were three bunkhouses built from 1911 to 1919 while the property was owned by Eudora Hull Spalding. The bunkhouses were built to house unmarried workers. Bunkhouse No. 1 was a large two-story building with a large living room, two dining rooms, a kitchen, a washroom, three bathrooms, a linen room, and 20 separate sleeping rooms on the second floor. Each room was furnished, and clean sheets and towels were provided twice a week. Below, in 1987, Catholic priest Fr. Ezequiel Mondragon prepares to give an outdoor Sunday church service at Rancho Sespe for the Mexican workers and families living on the rancho. (Both, courtesy of Ellen Frankenstein.)

Above is this 1960s Black and White Ball that was started in 1957. A group called Los Padrinos at St. Francis of Assisi Catholic Church created the fundraising event for the parish. The candidate who raised the most money was crowned queen of the event. From left to right are Linda Morales, Linda Reyes, Lupe Cervantez, Queen Dolores Gurrola, 1959 Queen Amelia Cardona, Rosie Frias, and Irene Alcozar. The Mexican community living at Rancho Sespe as well as in Fillmore thrived because their lives had meaning beyond work relationships with dominant groups. They fostered cultural continuity and spaces that became anchors. At right, George Espinoza plays his guitar at the 1946 Fillmore Frontier Days. George lived in nearby Santa Paula but participated in Fillmore's parades for many years, including during his retirement. He was known as the "strolling musician."

Above, Frank Munoz (in the front passenger seat), Fillmore's first Hispanic mayor, rides with council members Dewey Thompson and Delores Day (both in the back seat) in the 1973 May Festival parade. Below, parades were always a big event in Fillmore; large crowds lined the streets to watch the various local organizations and groups march or ride on floats. Below, a Mexican band called Banda Mexicana performs in a 1940s Fillmore parade.

One of Fillmore's first Mexican restaurants was Las Palmeras Café, located on Santa Clara Street. Although it was a small matchbox structure, it had a big reputation for good food. The restaurant catered to the Mexican community by offering home-cooked Mexican food. It was in business for many years. (Courtesy of Ellen Frankenstein.)

A 1932 photograph shows the interior of La Mexicana Market, one of the early Mexican markets in Fillmore. The store owner, Jose Gonzalez, is pictured in the center, and a store worker is at right. The photograph shows a meat section, where fresh meats were displayed.

Pictured is Abraham "Abe" Diaz Carreno. Originally from Oaxaca, Mexico, he settled in the valley in 1948. Abe owned and operated Abe's Hi-Way Market in Fillmore. He was an active member in the community and served as president of Fillmore's Rotary Club. (Courtesy of Ellen Frankenstein.)

A mid-1970s photograph shows Frank Munoz in retirement, still out and about in the city of Fillmore. Frank was a longtime resident of Fillmore, was active in the Mexican community, and was a beloved mayor of Fillmore. He had founded a concrete pipe works business in Santa Paula and in Fillmore on Santa Clara Street that was in operation for several years.

In this 1985 photograph, Fillmore mayor Ernie Morales holds a microphone as he addresses a demonstration against the English as the Official Language initiative. The political movement advocated the use of English in US government operations through the establishment of English as the only official language in the country. (Courtesy of Ellen Frankenstein.)

Here, people attend the 1985 demonstration against the English as the Official Language initiative. Some of the issues voiced by opponents of the initiative were the question of language as a right and language discrimination against bilingual people in violation of civil rights. To some, the initiative antagonized immigrants, and the creation of a national official language would only serve to discriminate, since the language initiative targeted specific groups of immigrants, particularly Spanish speakers. (Courtesy of Ellen Frankenstein.)

A farm laborer's house at Rancho Sespe, above, displays a United Farm Workers Association flag. Cesar Chavez was a farm laborer and lived for a period in the La Colonia area in nearby Oxnard. He supported the farmworkers' needs for better working and living conditions through organizing and negotiating contracts with farm owners. Chavez became a labor leader and civil rights activist. He cofounded the United Farm Workers and organized farmworkers to participate in strikes and boycotts. Below, supporters from Rancho Sespe march through the streets of Fillmore in support of the United Farm Workers' cause and efforts.

Four

Fillmore's Schools and Churches

Several students of various ages and a schoolteacher (right) are pictured in front of the first school in the Sespe area, located west of Sespe Creek on Sespe Road, now called Grand Avenue. Three districts were formed out of the western part of Sespe. The second school was San Cayetano, and the third was Fillmore.

An 1892 photograph shows students and teachers in front of San Cayetano School. After the railroad was built in 1887, the valley began growing citrus, and Mexican labor in the orange and lemon groves also grew. San Cayetano was the second school built in the district. It was called the "Mexican School" because it was the school most of the Mexican children attended.

An early photograph shows the second Cienega School, built in 1890. The first Cienega schoolhouse was located east of Fillmore and operated from 1873 to 1890. The early schools in the 1800s were typically one room with the teacher's desk on a platform and students sitting in rows facing the teacher.

This 1900 photograph shows the original Bardsdale School. The school had one room and a cloakroom. There were two entrances, one for girls and one for boys. An iron stove provided heat. Water had to be brought to the school, because there was no water on the school grounds.

The first Mountain View School is pictured in 1892. Mountain View School taught children from kindergarten to sixth grade. Mexican children living in the Sespe area were bused to Fillmore to attend Mountain View School.

Pictured is Fillmore's first high school. Hattie King built the school in 1909 and leased it to the school district. It was used as a high school for one year because a new high school opened on Central Avenue and First Street in 1910. King's building was later used as a residence.

Cactus Flats School and its students and teacher are seen in 1888. This was a temporary school while a school building was being constructed. The wood used to build the temporary school was loaned by local ranchers, who wanted the wood returned after the permanent building was completed. The wood was loaned on the condition that it not be cut or painted, resulting in its irregular shape.

A 1933 photograph shows Fillmore Union Grammar School's school bus. The valley was predominately agricultural, and many children lived in remote areas on farmland. The school bus was able to transport children who otherwise would have had a difficult time finding a way to school.

Mountain View School is seen in 1912 with building additions. Several students, mostly girls, and teachers are ready for May Day. The school was located in the 300 block on the east side of Mountain View Street. The girls and teachers are dressed in white, and the lead girl is carrying the American flag while two young boys man the flagpole.

Pictured is Fillmore Union High School in 1911. In 1924, the building became a junior high school because a larger high school was constructed next to it on Central Avenue and Second Street. Below, in the early 1930s, several female students pose next to cars in front of the old junior high school in celebration of an unknown, but seemingly special, girls-only event. Most of the girls are holding flowers. The junior high school building burned in 1937.

A 1938 photograph shows the new Fillmore Union High School. The school had to be demolished in the 1950s because it did not meet California's Field Act, which required school buildings to meet strict and specific earthquake safety standards. Below is Fillmore Union High School's first graduating class in 1911. From left to right are (first row) Albert Wiklund; (second row) Mary Cummings, Sarah King, and Mabel Arthur.

Fillmore Union High School's science building opened in 1938. The architecture and detail were characteristic of the 1920s and 1930s. A matching building was constructed on the north side of the Central Avenue two-story building. Many schools during that era were built with New Deal funds. The school's science department focused on preparing students for college and careers. Courses included biology, chemistry, physics, and integrated science.

Pictured are the 1915 Fillmore Union High School baseball champs in front of the high school. The high school has a sports hall of fame where the baseball champs are memorialized.

Several women and children attend an outdoor service at Hattie King's Sunday school in Bardsdale in 1915.

Several students pose for this 1904 photograph at the Bardsdale School. The school's teacher was Cecily Wilson Balden. The Bardsdale School had one of the earliest organizations called "Congress of Mothers," which was similar to the PTA. It was organized in 1925 and began when parents would meet with teachers. The mothers would bring lunches for the children, wash dishes after lunch, and raise money with cake sales to purchase utensils for cooking hot lunches for the children.

Trinity Episcopal Church is seen in 1935. Built in 1901 in Port Hueneme and owned by the Bard family, the building was no longer in use and was in need of repair. Mary Gerberding Bard donated the building to the Fillmore Episcopal congregation. The building was relocated 30 miles to Fillmore in 1933 and is still holding services to this day.

An 1898 photograph shows the Bardsdale Methodist Episcopal Church. Some of the first settlers in Bardsdale were German immigrants. A small German Evangelical church, which also served as a schoolhouse, was built around 1892. Church services were in German in the morning and English in the afternoon.

Sespe Methodist Church was on Grand Avenue. In 1889, the Pacific Improvement Company sold land to the Santa Paula Methodist Episcopal Church to build a church in the town of Sespe.

A 1912 photograph shows a Catholic church built on the corner of First Street and Central Avenue on property donated by Leon Hammond. The building was a simple wooden structure. By the mid-1970s, the congregation was in need of a larger church, and St. Francis of Assisi Church was built on Ventura Street. The original building became a private residence.

A different angle shows Bardsdale Methodist Church in 1927. The church was built in 1898 on Bardsdale Avenue in Bardsdale. Land for the church was donated by Thomas Bard, who also donated one of the stained-glass windows on the condition that the congregation donate the second window. The church is still an active Methodist church.

In 1910, Fillmore Methodists discussed the organization of the Fillmore Methodist Episcopal Church at Stephens Hall, located behind Stephens' Store, which today is La Estrella Market. The church was built in 1913 and had an active congregation until around 2014. Several members and children of the Fillmore Methodist Church are pictured in 1916.

The Women's Alliance of the Presbyterian Church is pictured above in 1903. The group sponsored educational, social, and enrichment activities. The women were homebodies who cooked and kept the home fires burning so that the men could farm or run businesses. They were religious and, like generations of women before and after them, met together for religious or social purposes. There were three Presbyterian churches in Fillmore through the years. The first was built in 1889 on Clay Street and Sespe Avenue and was destroyed by fire in 1912. The second church, built on the same corner, was also destroyed by fire. The third building was on the corner of Central Avenue and First Street. Below, the First Union Ladies Aid organization is pictured in 1906 at the home of Katrina (Kate) Burson in Bardsdale.

David Cook stands in front of the Piru Methodist Church. Cook was a publisher of religious tracts who had come to California from Elgin, Illinois, for his health in the 1880s. He financed and built the church in 1890.

The Church of the Nazarene is pictured on Central Avenue. The church's edifice was built in 1918 as the First Brethren Church. In 1978, the Church of Nazarene, located on Third Street, purchased the First Brethren Church building, and it has remained there ever since.

Pictured is the first Presbyterian church that was located on Sespe Avenue. The congregation was organized in 1888 at Kenney Grove. The building was later destroyed by fire.

The second Presbyterian church building was located on Sespe Avenue. The church relocated in 1929 to its new building, and the old building was sold to the Foursquare Church, which was destroyed by fire in 1939.

A 1912 photograph shows J.F. Martin (left), J.R. Moore (center), and D.P. Fulton in front of the Fillmore Methodist Church. Methodism in the valley parallels the history of the pioneers and their families. The church in Fillmore is no longer in operation.

The third Presbyterian church building in Fillmore is pictured in 1929, the year it was built. The church is no longer active. The building has become one of the city's landmarks.

Five

ENTERTAINMENT

Pictured is the Star Theater, which was located on Fillmore Street. Built by Wilmer Akers around 1900, it was the first real theater in town . It was a vaudeville house and silent movie palace. In 1902, Akers purchased a giant music box that played a 27-inch disc, which is now on display at the Fillmore Historical Museum.

Stearns Theatre on Central Avenue is pictured at left in 1927 with two men making repairs to the newly purchased theater. In 1916, Leon Hammond erected a commercial building where his home had previously stood. The building included the Towne Theater with a modern stage, a large door at the rear to bring scenery in and out, and a trapdoor on stage. It had an orchestra pit, which was later taken over by a large electric organ. There was also a five-foot-steep pitch to the seating area, which was unusual for the time. The theater had a large lobby with large picture frames where photographs of coming events were displayed. In addition to showing motion pictures, the theater hosted piano recitals, stage plays, public lectures, and "Country Store" nights where groceries were given away. Below is another photograph of the theater, which was sold in October 1926 to H.C. Stearns. The name of the theater was changed to Stearns Theater.

The above photograph shows the 1890 Fillmore High School (FHS) baseball team. From left to right are (first row) Earl Cole, Clint Howard, Earl Goodenough, and Jimmie Allee; (second row) Charles Haynes, Clyde Howard, ? Brockus, Rollie Martin, and Frank Howard. Sports became a big part of the community, and although FHS had only been in existence for a few years, the school's teams excelled in sports. The FHS baseball team won the county baseball championship in 1913–1914 and 1914–1915. Below is the 1915 FHS tennis team.

This town barbecue was held on July 4, 1900. The celebration of Independence Day became a big annual event when Congress established it as a holiday in 1870. The annual event was always celebrated with a barbecue, parade, picnic, games, and other festivities.

In 1895, Fillmore pioneer families are on a camping trip. Pictured, not in order, are Osbourne S. Bookhout, Dr. J.P. Hinckley, Elizabeth Phillips (mother of Martha Phillips Elkins), Ella Tietsort, William W. Elkins, Squire Tietsort, Martha Phillips Elkins, C.C. Elkins, and children. Osbourne Bookhout and his wife met Dr. John Hinckley and his wife when they were living in Tulare, South Dakota. They came west with the Hinckleys and made Fillmore their home.

A 1927 stage musical called *Buddies* featured high school students and a musical group dressed in military uniforms depicting World War I soldiers. The girls are dressed in European costumes. The play was produced by the Fillmore Veteran's Service Club and was standing room only at the high school auditorium. Stage musicals were a big source of entertainment in Fillmore during the early 1900s.

Francis Osborn's dance class in 1958 shows an all-girls ballet group. Osborn's dance school arranged many youth performances, including singers and dancers. Classes also taught young students confidence, concentration skills, and an understanding of music and rhythm. The physical aspect of ballet and dance strengthened young bodies and minds.

Above, the Harmonson and Rood Orchestra plays for Fillmore's 1906 Fourth of July outdoor celebration. The musical group sometimes had up to 12 musicians—mostly members of the Rood and Harmonson families of Fillmore. Below, the Harmonson and Rood Orchestra played for dances at Stephens Hall on Sunday nights and also played for a local church on Sunday mornings. Musical groups were popular in town, including the Fillmore City Band and the Fillmore Citrus Association's Mexican Band.

In 1918, sisters Edith (left) and Alice (center) Moore and a friend—all dressed in bathing suits and bathing caps—enjoyed the outdoors at Swallows' Nest, a popular swimming hole on Sespe Creek north of Fillmore.

Four thousand people attended the celebration of the opening of the Bardsdale Bridge on November 23, 1909. Crossing the Santa Clara River was a major problem until the bridge was built. High water and heavy storms often prevented people from reaching Fillmore and Bardsdale. Everyone at the celebration received a button that proclaimed "All Roads lead to Fillmore. Meet me there."

At a 1930 hunting trip campsite, one man sits relaxing with a pipe and the other cooks up a meal. Hunting parties and camps in the Agua Blanca area were popular. Below, George Henley's fishing camp is seen in 1910. The Sespe Creek was noted for its native steelhead trout. During fishing season, particularly May 1, which marked the opening of trout season, the town of Fillmore bustled with fishermen. No rooms were left to rent, restaurants were full, and one would have to "take your own rock to stand on." George Henley is second to left.

A late-1800s photograph shows L.W. Fansler with the violin, daughter Goldie Fansler at the piano, Glen Fansler standing, and Dora Combs Fansler seated. Families often gathered in their living room and created their own entertainment. Electricity came to Fillmore in 1907, and there was no television or radio in the homes. Playing a musical instrument such as piano or violin was almost a required skill.

Pictured in the early 1920s is Leon Harthorn, son of Judge C.W. Harthorn who established the Cash Commercial store, driving the Harthorn's Cash Commercial Co. No. 2 truck as an entry in a Fillmore parade. The decorated truck appears to have a Native American theme, with a teepee on the bed of the truck.

The Fillmore High School band is pictured at the May Festival Parade in 1940. Young people can be seen on top of the buildings to watch the parade. The May Festival was a community affair sponsored by the Fillmore Board of Trade. It was one of the most popular and well-attended events of the year.

Festival queen Ruth Johnson and her court are seen in 1946 during Fillmore's Frontier Festival. The festival began with a parade, and the winner of the most votes became the festival queen with four runners-up as her court. Tickets were obtained from local merchants: one ticket for each dollar's worth of merchandise, or they sold for 10¢ a ticket or three tickets for 25¢.

Children pose for this 1921 maypole dance event at Mountain View School. The girl in the middle is the maypole queen. She is wearing a crown and holding a stick to resemble a scepter. A maypole is a tall wooden pole that is used at various European folk festivals. Since many of the early pioneers of the valley were German, the maypole tradition carried over.

A 1910 photograph shows young Bardsdale Cadets. Cadets often drilled twice a week on schoolgrounds wearing uniforms and carrying dummy rifles or sticks. Drills included marching in formation, participating in mock battles, and shooting practice. Pictured here are, in no particular order, Alfred Clayton, the LeBard boys, Tom Cruson, and other cadets from Bardsdale.

The 1947 Frontier Days parade moves down Main Street. A car pulls a float with young women in bathing suits, including one holding on to a water-ski rope. In back of the float are people on horseback. People stand in the street next to parked cars that line both sides of the street to get a good view of the parade.

Pictured are the 1938 Fillmore Festival queen and her court. Mildred Baum Legan won the most votes from tickets sold from local stores. From left to right are Laura Davis Smith, Mildred Baum Legan, and Vernice Jones Miller.

A 1952 parade on Central Avenue shows the Fillmore High School marching band. Crowds of people line the sidewalk to watch. During the war years, the festival was suspended; the active Junior Chamber of Commerce revived the Fillmore Festival in 1946. The Jaycees brought screen and television personalities as grand marshals and appointed judges on various activities. At right, Dr. Bill Manning (left) and Sheriff Howard Durley (center) are on horseback followed by American Legion Post No. 481 in the 1952 parade. Dr. Manning practiced in Fillmore from 1912 to 1940.

Jim's Pool Hall, to the left of Ballard Furniture on the west side of Central Avenue, is pictured in the 1920s. Jim Ipswitch purchased the pool hall, which was known as the Fillmore Billiard Parlor, in 1923. Jim's son, Jack Ipswitch, and a partner purchased the pool hall from Jim in 1940. Although the billiard parlor's name was changed to the Amusement Parlor, it still remained a pool hall.

Residents of Fillmore had a surprise snowfall on January 8, 1949, when the temperature dropped to 28 degrees and remained low for almost a week. Citrus growers kept smudge pots going to save the navel orange crop. Schools and businesses closed, possibly for some snow fun. Children and adults hopped on makeshift sleds, and some put on skis. Pictured are Phil (left) and Fred Young.

Six

Profiles of Pioneer Families

The Wileman family is pictured in Bardsdale in 1898. The family moved to Fillmore in 1895. From left to right are Isamiah (Felty) Wileman, Ella and Ross on the mule, Bruce, Thomas, William, and Harry. Early pioneers were the catalyst to California's transformation from a rancho economy to an agricultural economy. Agriculture became one of the state's main sources of wealth.

Pictured is Ray Ealy in his horse buggy in 1910. Originally from Iowa, Rush Ealy settled with his wife, Ella, and sons Ray and Fred just east of Fillmore in the 1880s. He built a store and stagecoach stop and called it Cienega. Rush also served as postmaster at Cienega from 1884 to 1889.

C.C. Elkins (left) and McCoy Pyle are pictured in the mid-1800s. Elkins came to Fillmore in 1877 and opened a store and olive oil factory. Pyle was killed while serving as a constable. McCoy Pyle's brother, Everitt Pyle, was Fillmore's second mayor.

Dr. John Powell Hinckley and his wife, Cora, pose in front of their home in Fillmore. Dr. Hinckley and Cora settled in Fillmore in 1889 with their children, Ira, Eugenia, and Mary. John went to medical school at the University of Vermont in 1876. He first set up practice in Tulare, South Dakota, but relocated in 1893 to Ventura County, where his fourth child, Vinnie, was born.

Dr. Ira Hinckley, son of Dr. John Powell Hinckley, and Ira's wife, Catherine Cruson Hinckley, pose for this photograph with their children, Hattie Mae (left) and Lawrence, in the early 1900s. Ira Hinckley graduated from Berkeley School of Dentistry and first practiced in Ventura until he moved to Fillmore in the early 1900s. Ira was one of Fillmore's early dentists and pharmacists.

An early-1920s photograph shows Kaichiro "John" Inadomi, who emigrated from Japan to the United States in 1914 and settled in Fillmore in 1925 with his wife, Mitsuyo Ogawa. John worked for the Limoneria Ranch in Santa Paula and became foreman at the age of 22. He later moved to Fillmore to start a business. Pictured below are John and his family. From left to right are (first row) John, baby Yoshiharu on John's lap, Mitsuyo, and Midori; (second row) Kumanosuke, Monda, and Wari. In 1942, during World War II, President Roosevelt issued Executive Order 9066, which called for the forced removal and incarceration of over 125,000 individuals of Japanese ancestry, including Japanese American citizens. The forced removal of Japanese to isolated internment camps caused many to lose their homes and businesses.

John opened the Inadomi Department Store on Main Street in 1925 with the help of a Japanese attorney, Katsutaro Tanigoshi. The alien land laws in the United States prohibited John from purchasing property because he was not an American citizen. Tanigoshi purchased the property through his Caucasian American wife's name and then resold the property to John's American-born children, who were both toddlers at the time. John's store sold groceries and other items and catered to the Latino community. John became well known in the community and was a member of the Presbyterian church and Rotary Club. Before John built his house, the family lived in a shack behind the store. Below, John's house was built in 1939 beside the store on Main Street, three years before the Japanese internment.

Charles Whitney Harthorn is pictured in 1924. A former sea captain, he and his family came from Thomaston, Maine, and settled in the Buckhorn area in 1901. Harthorn bought Richard Stephens's store with his partner, W.G. Cornelius, and renamed the store Cash Commercial Grocery Store, selling dry goods and groceries. He later bought out his partner and discontinued selling dry goods to focus on his customers' grocery needs. Below is the interior of Cash Commercial in 1920, when it sold dry goods and groceries. Two employees behind the counter, W.G. Cornelius (left) and Charles Harthorn (right), are waiting on a female customer with her child.

An early-1900s photograph shows Leon Hammond and his wife, Trinidad Robles. Leon was born in Brazil in 1866, and Trinidad was from Santa Paula. They settled in Fillmore around 1905. Leon built the Towne Theater on their original homesite on Central Avenue. He also donated the land for the first Catholic church on Central Avenue.

John Pointz Trotter, pictured in the 1890s, was from Missouri and homesteaded in Pole Canyon in the early 1890s. He later sold the homestead and moved to Fillmore, where he opened the town's first livery stable. John helped found the Fillmore State Bank and was director of the Fillmore Improvement Company.

Rev. John Guiberson settled in the valley in 1869 and preached at the Methodist church in Cienega. The first Sunday school in the Santa Clara River valley was near Cienega, initially a stage stop on a wagon road leading to the coast. Guiberson led services at the first celebration of the nation's founding on July 4, 1876. Many of Fillmore's early churches held services outside under oak trees, and Sespe is regarded as the birthplace of churches in the east end of Ventura County. Below are Samuel Guiberson and his family. Samuel was Rev. John Guiberson's son. Samuel became a Ventura County supervisor.

William Shiells and his family are pictured in the early 1900s. Shiells emigrated from Scotland to the United States in 1872. He and his brother, James, purchased 1,200 acres of oil-rich land south of Fillmore in 1884, which they leased to the Montebello Oil Company in 1910. The land became one of the most productive oil leases in the area.

Pictured are Cyrus Kenney and family in the late 1800s. Cyrus and his wife, Elvira, came to Sespe in the 1870s. Although the Kenney family did not own Kenney Grove, a favorite place for picnics, the site was named after the family. From left to right in back are Maude (Harry Kenny's wife), Mabel, Hazel Verrue, Elvira, and Cyrus, with Cyrus and Elvira's grandchildren in front.

The Sanitary Dairy delivery truck had solid rubber tires in 1925. Clifford Hardison moved to Ventura County from Maine in 1906 and relocated to Sespe in 1911. Clifford and his wife, Elvira, founded the Sanitary Dairy in 1916. The name meant that the cows were certified as healthy and the raw milk was handled according to regulations. The dairy prospered and expanded for many years. Home delivery demand decreased, and the delivery routes were discontinued in 1977. The business closed in the 1980s. The dairy was declared a county landmark by the Ventura County Board of Supervisors in 1989. From left to right are Cliff, Dorothy, Russell, and Evelyn Hardison. At left is Elvira Hardison in 1955. Elvira added a stand alongside the dairy and offered fresh-squeezed orange juice, dairy products, sandwiches, candies, sodas, and other items.

George King and wife Hattie Bussick are pictured in the late 1800s. George worked for David C. Cook, a wealthy publisher of religious tracts who is regarded as the founder of Piru. Cook laid out the town of Piru and built a rail line to serve the town. George and Hattie later moved to Bardsdale, where he became the property manager for Thomas Bard, a wealthy oil baron and cofounder and president of Union Oil Company. Bard was the nephew of Thomas A. Scott, a Pennsylvania railroad and oil baron who purchased Rancho Ojai in 1864 and owned several properties in the county. Below, the home of George and Hattie in Bardsdale was built around 1929.

Jacob Michel and wife Matilda came from Indiana in the 1890s and settled in Bardsdale. They were one of the early pioneer families in the valley. Jacob and Matilda had seven children, Muriel, Lloyd, Mary (Mayme), Clarence, Charles, Irvine, and Ted. All the children were born and reared in Bardsdale. Below, many of the residents of Bardsdale in the late 1800s are seen from the Bardsdale hills looking north toward the Santa Clara River. The early settlers were mostly divided between German and English-speaking people.

The Carlo Basolo family emigrated from Italy to the United States in 1889 and settled in Bardsdale. Carlo had 10 children. Below is a 1914 photograph of a Jenny airplane on an airstrip that Carlo created on his ranch. The Jenny was one of a series of biplanes built by the Curtiss Aeroplane Company of Hammondsport, New York, which later became the Curtiss Aeroplane and Motor Company. The planes were originally produced as training aircraft for the US Army but after World War I were used as civilian aircraft.

German immigrant Rudolph Haase and his family pose in the 1920s. Haase came to Bardsdale in 1889. He helped bring water to the town, which significantly impacted farming success there. He was instrumental in the formation of the Southside Improvement Company, which brought water from the Santa Clara River to Bardsdale. From left to right are (first row) Alma; (second row) Rudolph, Hilda, and Emilie (Rudolph's wife); (third row) Otto, Jean (Herman's wife), Albert, Herman, and Elsie. Below is the Haase family on a bus to San Diego in 1915. One pennant reads "On the Boundary Line" and the other "U.S. & Mex." San Diego is located on the US border with Mexico.

Thomas "Tommy" Arundell and family are pictured around 1910. Tommy began beekeeping at Pole Creek in 1879. He had as many as 700 stands, which are also known as beehives. The stands are bases for elevating and securing the beehives. From left to right are (first row) Arthur, Thomas, Ernest, Inez (Tommy's wife), and Allen; (second row) Louise Arundell Phillips, Frank, Norman, and Elizabeth.

Pictured is the Arundell family's secluded and remote adobe in Pole Creek Canyon. In addition to beekeeping, Tommy Arundell also farmed apricots. Apricots were widely grown in the valley before citrus. Apricots were introduced in Ventura County by the San Buenaventura Mission and spread throughout the county by the settlers.

The late-1800s photograph at left shows sisters Mahala Azbell Stone (left) and Angeline Azbell Baum. The sisters came to the valley in 1871, and both lost children to the diphtheria epidemic in 1878. Children were widely affected, and many lost their lives. Families became sick, and some parents lost all their children to the disease. Mahala's three older sons survived, as did daughter Ellen; daughter Ada was born after the epidemic. Below is a 1910 photograph of Mahala Stone taken in front of her house on Central Avenue. The house was located on the west side of the 300 block. She had the home moved to Mountain View Street in 1911 when Central Avenue became more of a commercial street. She lived in the house on Mountain View Street with her sons until her death in 1933 at the age of 95.

Seven

FILLMORE TODAY

Fillmore's "F" still sits on a hill above the city. Built in 1936, it has become a landmark and a part of the cultural landscape, which is pride in one's community and civic identity. Solar lights illuminate the letter at night, thanks to John Heilman and friends, who installed them, and the Fillmore High Alumni Association, which keeps a fund going to maintain the lights.

Pictured in 2022 is Fillmore City Hall, which was built in 1996, two years after the Northridge earthquake damaged and destroyed several buildings in Fillmore. The new city hall's neoclassical-style architecture, with six columns and six arched windows at its front entrance, depicts grandeur of scale yet remains simple. Adjoining city hall is Central Park Plaza (left), a quaint city park with a three-tier fountain in the center. Its picturesque landscape makes it a popular place for a stroll or picnic, and it is frequently used for public and private events, such as festivals, weddings, and celebrations.

Fillmore's four-face town clock is on Central Avenue. The Fillmore Rotary Club supported the vision of a town clock and donated $5,000 for the project. The clock was built by the Verdin Company and installed in 2009 as part of the Central Avenue Storm Drain Project.

Constructed in 1917 on the corner of Central Avenue and Main Street, this building first housed the Fillmore State Bank. In 1927, it became a branch of the Bank of Italy, which changed its name to Bank of America in 1930. In 1965, it became a branch of Bank of A. Levy. Today, it is owned by the Diamond Realty agency. The building is listed in the Ventura County Historical Landmarks series.

Parkview Court senior apartments were built in 1976 on the southwest corner of Central Avenue and Main Street. The site was once the Sparr Packinghouse, one of Fillmore's earliest packinghouses. The packinghouse burned in 1913 and had to be rebuilt. It suffered damage from fire again in 1970, long after it had closed its doors.

Estrella Market is on the west corner of Central Avenue and Main Street. The building was completed in 1910 and was called Stephens' Store. The Fillmore Post Office shared the Stephens' Store building but was separated from the store by a connecting door and had a front entrance on Central Avenue. Richard Stephens was the postmaster for over 16 years. In 1911, Stephens' Store became the Cash Commercial store.

The Fillmore Post Office is seen on Central Avenue. Philippe P. Roche was appointed postmaster in 1915. The post office moved some time in the early 1900s from the Stephens' Store building to the east side of Central Avenue in a jewelry store owned by Roche. The post office moved again in 1920 to the Masonic temple, built in 1919, and then to its final location in 1951.

This photograph shows the east side of Central Avenue and Sespe Avenue. Patterson Hardware is one of the oldest businesses in downtown Fillmore. It was originally one of five stores owned by three Hickey brothers of Ventura. The Central Avenue store was the first location, and Harvey S. Patterson was the manager. Patterson purchased the business from the Hickey brothers in 1937 and changed the name to Patterson Hardware.

Looking down Central Avenue toward the Santa Clara River, which runs below the mountains, downtown Fillmore on Central Avenue comprises two blocks where most businesses are located. The city clock sits at the railroad crossing. Below is the Santa Clara River Valley Railroad Historical Society's Railroad Visitor Center in Fillmore. The center is on Main Street adjacent to the railroad tracks. It displays historical artifacts, photographs, and information relating to the railroad in Fillmore and has a model train on display. It is a treasure trove of information on how the railroad changed the Santa Clara River valley and how the town of Fillmore was born with the coming of the railroad.

Daniel Haro is pictured at right in front of the original Fillmore train depot, which was built in 1887, when the Southern Pacific Railroad completed its line from Saugus to Ventura. The depot was the hub of the town. The train transported passengers, materials, and agricultural products. The last passenger train came through Fillmore on January 13, 1935. In 1974, Southern Pacific closed the depot, and the track between Piru and Newhall was later removed. On March 3, 1974, local teacher and author Edith Jarrett purchased the depot for $1.05 (the 5¢ was tax) and paid to move it across the street from its original location, where it became the site of the Fillmore Historical Museum. Below, train tracks run adjacent to the site where the train depot building now sits.

The Fillmore Unified School District Building is on Sespe Avenue. By early 1889, the school district split into three: Sespe, San Cayetano, and Fillmore. The district is an integral part of the city and is the city's single largest employer.

The Boys & Girls Club and Fillmore Community Center is on First Street. Boys & Girls Clubs of America was founded in 1860 on the concept that "boys who roamed the streets should have a positive alternative." The club here was originally founded as a Boys Club in 1949 by the Santa Paula Police Department. Girls were eligible to become members in 1968. The Fillmore club is a branch of the Boy & Girls Club of Santa Clara Valley for youths from kindergarten through 12th grade.

Here is the Fillmore Library as seen today. The Fillmore branch library has been in existence for over 100 years. It began as a traveling library, then in 1916 was housed in the Ventura County Co-op building on Central Avenue. In 1917, it moved to the second floor of the then-new Fillmore State Bank building. After 1917, the library relocated several times until its current building was erected in 1955. In 2022, the library underwent expansion and renovation.

The Fillmore-Piru Veterans Memorial Building is on Second Street. The building sits on land donated by the Shiells family and was built in 1952 for the purpose of recognizing local veterans. It provides support for veterans of all US armed forces by connecting veterans with government agencies and nonprofit organizations at the federal, state, and local levels. The building is also used for church services, events, and celebrations.

Trinity Episcopal Church, also called the "Chocolate Church" because of its rich brown color, is on the corner of Saratoga and Second Streets in a residential area of the city. It was built in 1901 by Sen. Thomas Bard for his wife in Hueneme, but the building was moved to Fillmore in 1933. It is still an active congregation.

Pictured here is the former Fillmore Presbyterian Church on Central Avenue. Built in 1929, it has become one of the city's largest landmarks, taking up an entire block between First Street and Kensington Drive. It is now an event center that is used for special events such as weddings.

People gather in front of the Fillmore Town Theatre for a classic car event held on Central Avenue in 2022. The single-screen theater was built in 1916 and was a former vaudeville and movie theater. After being closed for several years, its grand opening in October 2022 featured a free screening of *...And the Earth Did Not Swallow Him*. The screening is dedicated to the senior migrant agricultural workers of Fillmore. Pictured at center is Daniel A. Haro.

The headquarters building of the *Fillmore Gazette* is on Sespe Avenue. The *Gazette* is the official newspaper of record for the city of Fillmore. The company is owned by Martin Farrell, who is also the editor of the weekly paper, which covers local news, sports, business, politics, and community events. Preceding the *Gazette* was the *Fillmore Herald* (1907–2006), serving the town even before its incorporation in 1914.

Giessinger Winery & Tasting Room is on Santa Clara Street adjacent to the Fillmore Historical Museum alongside the train tracks. The building was once a welding shop. Owner Edouard Giessinger, a wine hobbyist, purchased the old building and opened the winery and tasting room. It has been operating for more than two decades.

When Balden Towne Plaza broke ground in 1995 and Vons supermarket opened there in 1996, it became Fillmore's premier and most prominent neighborhood shopping center. Today, Rite Aid, Baskin & Robbins ice cream store, eateries, and a nail salon, to name a few businesses, make up only a part of the shopping center. Others that share the Vons parking lot include Starbucks, a barbecue restaurant, Chinese fast food, Subway, Wing Stop, and Jamba Juice.

The Fillmore Police and Sheriff's Office is on Sespe Avenue in what was formerly city hall. Law enforcement in the town began in 1872, when Ventura County separated from Santa Barbara County. By 1897, Owen Miller was Fillmore's first constable. In 1910, John "Jack" Casner was elected constable, and he served for 31 years. In 1922, Earl Hume was hired as a traffic officer, and he became chief of police in 1925, serving Fillmore for 42 years. In 1987, the city contracted law enforcement services with the Ventura County Sheriff's Office, supplemented by volunteer Explorer Scouts, a search and rescue unit, and a citizen patrol. Below, Fillmore City Fire Department No. 91, on Sespe Avenue, was founded in 1914 as a volunteer force. In 1945, an agreement with Ventura County allowed for one fire truck and one man to be assigned to the city fire station. Station No. 27, on C Street, was completed in 2019 and serves the unincorporated areas near and around Fillmore in cooperation with Fillmore's city fire department. (Below, photograph by Sebastian Ramirez.)

Fillmore's Two Rivers Park is a 22-acre park at the corner of C and River Streets. It opened in 2009 and was named for its location near the Santa Clara River and Sespe Creek. It is one of the city's largest parks and offers baseball and soccer fields, a playground, a walking and bicycle path, a skate park, and a dog park. At left, a bicycle and walking path runs parallel to the Santa Clara River at the new Bridges housing development in Heritage Grove. The development includes a 27-acre nature park, walking trails, a basketball court, a children's playground, and a picnic area.

The entrance to the Bridges is off Highway 126 and Mountain View Street. In 1928, the Bardsdale Bridge superstructure was destroyed when the St. Francis Dam failed and devastated the valley. In 1938, another flood washed away the north approach to the bridge. In 1993, the structure was declared too narrow and risky to cross. The bridge was replaced with a modern structure in 1994, and the green 1928 bridge superstructure was purchased by the city and moved to a storage location on the north side of the river near Pole Creek. The bridge remained in storage for years until it was repurposed as the Mountain View Street entrance into the Bridges. The first phase of the Bridges development began in 2018 and consisted of 125 homes and Rio Vista Elementary School. The second phase consisted of 166 houses and a park. The third—and last—phase will add 459 additional houses.